Alan McKirdy has written many popular books and book chapters on geology and related topics and has helped to promote the study of environmental geology in Scotland. His other books with Birlinn include *Set in Stone: The Geology and Landscapes of Scotland* and *Land of Mountain and Flood*, which was nominated for the Saltire Research Book of the Year award. Before his retirement, he was Head of Knowledge and Information Management at Scottish Natural Heritage. Alan is now a freelance writer and has given many talks on Scottish geology and landscapes at book festivals and other events across the country.

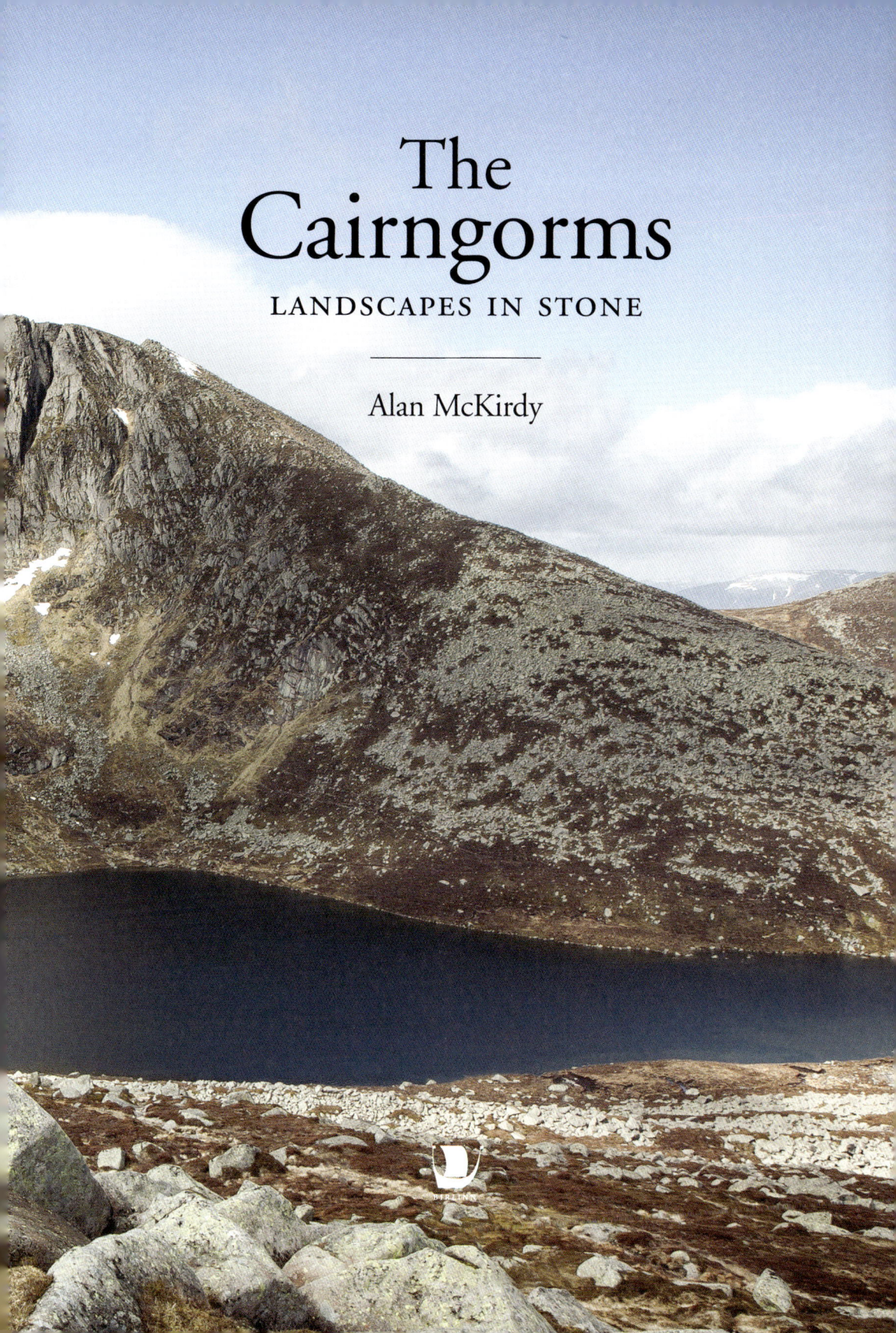

For Nigel Trewin

First published in Great Britain in 2017 by
Birlinn Ltd
West Newington House
10 Newington Road
Edinburgh
EH9 1QS

www.birlinn.co.uk

ISBN: 978 1 78027 370 9

Copyright © Alan McKirdy 2017

The right of Alan McKirdy to be identified as the author of this work has been asserted by him in accordance with the Copyright, Designs and Patents Act, 1988

All rights reserved. No part of this publication may be reproduced, stored, or transmitted in any form, or by any means, electronic, mechanical or photocopying, recording or otherwise, without the express written permission of the publisher.

British Library Cataloguing-in-Publication Data
A catalogue record for this book is available on request from the British Library

Designed and typeset by Mark Blackadder

FRONTISPIECE.
Lochnagar.

Printed and bound in Britain by Latimer Trend, Plymouth

Contents

	Introduction	7
	The Cairngorms through time	8
	Geological map	10
1.	Continents collide	11
2.	Devonian deserts	17
3.	Mind the gap!	20
4.	The big freeze	22
5.	The nature of the Cairngorms	32
6.	A natural playground	37
7.	Places to visit	42
	Acknowledgements and picture credits	48

Introduction

The geology of the Cairngorms was created on a timeline that stretches far back into the mists of the dim and distant past. Much of the Cairngorms are underlain by granite that formed deep within the Earth's crust and 'surfaced' as the overlying layers of rock were stripped away by the forces of erosion – ice, wind and water.

The geologically recent Ice Age re-engineered the Cairngorm landscape in dramatic fashion. The Lairig Ghru, an incised glen that runs north-west to south-east, is one of the most prominent landmarks in the Cairngorms. This iconic landform was carved by a fast-moving stream of ice that cut deep into the granite massif. The bedrock is hard and, although the area has been heavily glaciated, the Cairngorms boast 20 Munros, the highest of Scotland's peaks. In fact, four of the five highest mountains in the country are in the Cairngorm range.

The area attracts climbers, walkers and assorted adventurers who want to pit themselves against some of the most challenging conditions to be found anywhere in the UK. The plants and animals of the high plateau also need to be hardy to survive the severe winter conditions. The higher reaches of the mountains are rich in montane vegetation such as lichen heath and other habitats that support many rare species.

Opposite. Corrie Lairige, Spittal of Glenshee.

The Cairngorms through time

Period of geological time	Millions of years ago	Scotland's global position	Environments and events in the Cairngorms
Anthropocene	Last 10,000 years	57° N	The barren landscape, abandoned by the vanishing cover of ice, was initially clothed by pioneer plant species and later by the spread of pine forest. The Great Wood of Caledon eventually covered the area. Early settlers then felled much of the tree cover. Subsequent management of the land by sporting estates has maintained a predominantly heather moorland ecosystem and extensive forest cover.
Quaternary	Started 2 million years ago	Present position of 57° N	This was the age of ice. Many advances and retreats of the ice are recorded during this period. Glaciers created the familiar landscapes of today. • 12,900 to 11,500 years ago – the climate became cold again as another advance of the ice took place. Small glaciers developed and the land was gripped by freezing conditions. • 15,000 to 12,900 years ago – during this time, there was a rise in temperatures and much of the ice cover melted. • 18,000 to 15,000 years ago – a slight warming of temperatures led to the formation of meltwater channels, kame terraces, eskers, kettleholes and ice-dammed lakes. • 30,000 to 18,000 years ago – the ice sheet expanded in extent, reaching a maximum about 22,000 years ago. • 2.6 million to 30,000 years ago – throughout this period, the climate changed from warm to cooler conditions. The extent and thickness of ice cover was driven by these climatic changes. Major landscape features, such as corries and glacial troughs (e.g. the Lairig Ghru) were gouged out of the bedrock, and distinctive tors formed.
Neogene	2–24	55° N	The climate cooled as the Ice Age approached.
Palaeogene	24–65	50° N	The tropical climate deeply weathered the granite bedrock, allowing tors to form at a later time, and the Cairngorm plateau surface was shaped.
Cretaceous	65–142	40° N	The land that was to become the Cairngorms was one of two areas of land that emerged from a warm tropical sea, as the world's oceans rose to new heights.

Period of geological time	Millions of years ago	Scotland's global position	Environments and events in the Cairngorms
Jurassic	142–205	35° N	Sea levels rose worldwide, but the Cairngorms remained above the waves. No deposits of this age are recorded in the Cairngorms.
Triassic	205–248	30° N	The land was arid with significant rainfall at times, but no rocks of this age are recorded in the Cairngorms.
Permian	248–290	20° N	Desert conditions prevailed, but no deposits of this age are recorded in the Cairngorms.
Carboniferous	290–354	On the Equator	Tropical forests covered what was to become the central belt of Scotland, but no deposits of this age are recorded in the Cairngorms.
Devonian	354–417	10° S	The Caledonian mountains were eroded from Himalayan heights to gentler contours. Sandstone deposits of Devonian age are recorded near Tomintoul.
Silurian	417–443	15° S	Final events unfolded in relation to the closing of the Iapetus Ocean as 'Scotland' and 'England' were finally united.
Ordovician	443–495	20° S	The Iapetus Ocean reached its widest point. The land that was to become Cairngorms lay at the northern margin of this ocean. Rocks from these times were laid down in the ocean deeps as sands, limestones and muds. The continental collision started in late Ordovician times to create the Caledonian Mountains.
Cambrian	495–545	30° S	
Proterozoic	545–2,500	Close to South Pole	The sediments that were cooked and squashed as the Caledonian Mountains were formed were laid down in the Iapetus Ocean at this time. They are known as the Dalradian.
Archaean	Prior to 2,500	Unknown	No rocks of this age are recorded in the Cairngorms. The Earth was formed 4,540,000,000 years ago.

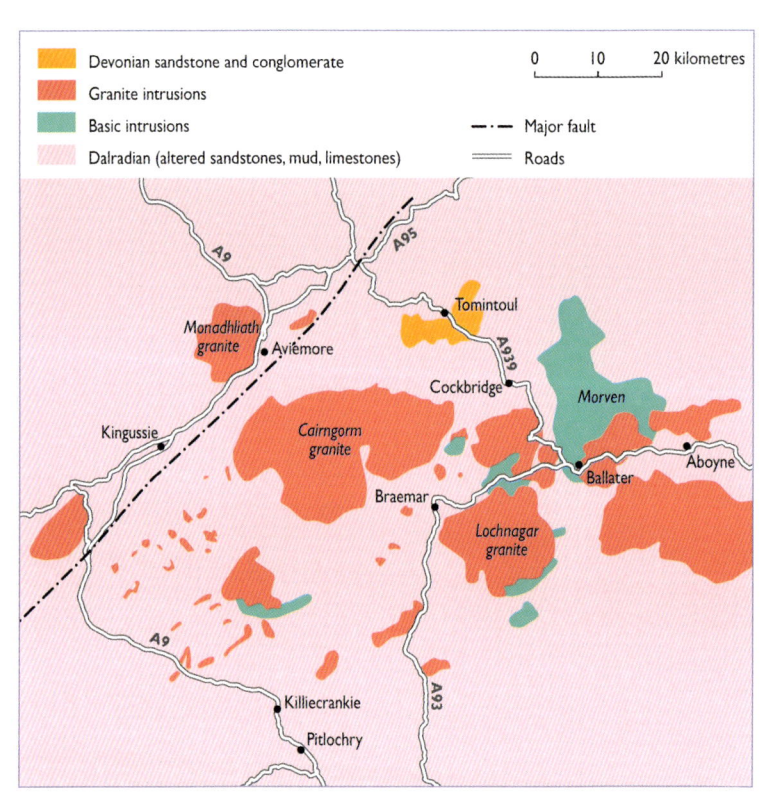

Geological map of the Cairngorms. Compared with other areas of Scotland, the geological map of the Cairngorms is simplicity itself. The bedrock consists of just three elements: metamorphic rocks (altered by heat or pressure) of the sequence known as the Dalradian; the many patches of granites and related rocks that form the foundation of the most dramatic peaks of the area; and a small patch of sandstones and conglomerates of Devonian age that encircle Tomintoul. But on this simple patchwork of bedrock are laid a complex array of landforms that were developed during the last Ice Age – a period of just over two million years. In places, thick layers of sand and gravel were dumped by the retreating ice as temperatures rose at the end of the Ice Age. Prior to that, when the ice covered the very highest peaks, its erosive power carved great corries and glens that are the signature views of the Cairngorms today. The evolution of the landscape continues as debris flows on the very highest ground, softening the profile of the slopes. On the lower ground, the Rivers Feshie, Avon and Dee periodically inundate the adjacent floodplains with dramatic effect. This is still an active and evolving landscape that demands respect and careful management.

1
Continents collide

Continents have been on the move across the face of Planet Earth over the last three billion years. The one on which Scotland now stands is no exception. In much earlier times before life on Earth was established, the land that was to become Scotland was located close to the South Pole.

Since that time, it has been propelled by forces that originate far beneath our feet to a location some 57° N of the Equator. To discover the driving force for this movement, we need to look deep within the Earth itself. The surface of the Earth is divided into a series of plates that are moved around by powerful currents in the layer called the Earth's mantle, located just beneath the Earth's crust, the outer layer. The Earth's core has a raging temperature of 6,000°C and heat leaks into the overlying mantle to set up a convection motion that drives the overlying plates at a slow but relentless pace of, on average, 3cm per year.

Over many hundreds of millions of years, the way in which land and sea are distributed around the globe has changed remarkably. Oceans have come and gone and, as continents collide, so the layers of sands and muds, laid down on the ocean floors, have been squeezed and squashed to form mountains. Many of the highest peaks in

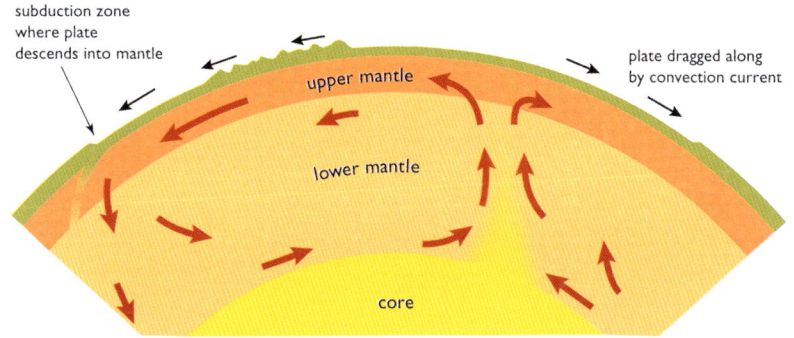

Section through the Earth showing core, mantle and crust. The crust is the outer layer, shown in green.

Scotland were formed in this way, including the Cairngorms.

So let's rewind to the time when the oldest rocks in Scotland were part of a larger continental landmass called Laurentia, which also included North America and Greenland. An ocean as wide as the present-day Atlantic Ocean separated Laurentia from another landmass to the south called Avalonia and one to the east known as Baltica. That ocean was called the Iapetus Ocean. It existed for around 250 million years and it left an indelible mark on the Earth's surface.

The time-lapse series of views (right) represents the closure of the Iapetus Ocean. Thick deposits of sand, mud, limestone and lava had built up on the ocean floor as the Iapetus Ocean reached its widest point. Around 500 million years ago, the ocean began to close. A slab of ocean floor sank to be consumed as it headed downwards into the mantle – a process known as subduction. As it did so, the descending plate and the rocks above melted, and this molten liquid rock returned to the surface, creating a chain of volcanic islands. The process of ocean closure continued. The sands and muds on the ocean floor were squeezed and squashed as the continental landmasses moved ever closer.

It was the collision of the volcanic islands with Laurentia that did the real damage, causing the layers of ocean floor sediments to be looped and bent into a contorted pile of 'metamorphosed' or altered

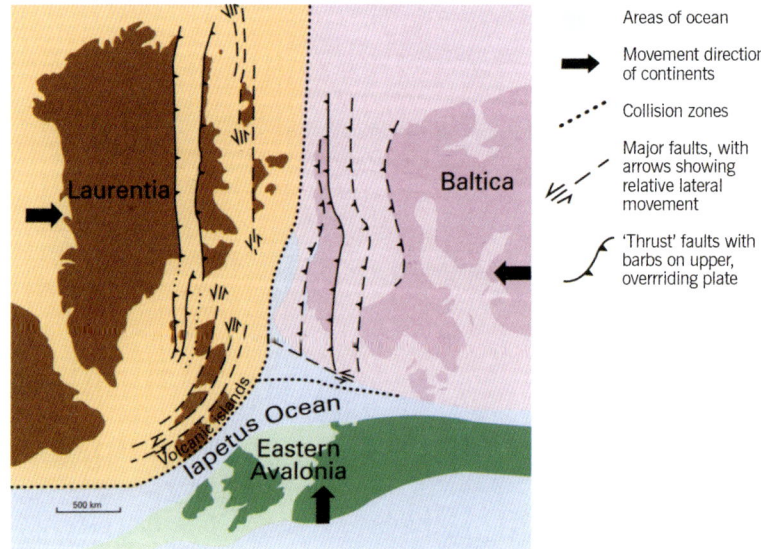

As the continents on either side of the Iapetus Ocean came together, the ocean disappeared. The layers of mud, sand, limestone and lava that had accumulated on the ocean floor were buckled and folded as the slow-motion 'continental car-crash' unfolded. The result was a mountain chain the height of the present-day Himalayas that stretched for many thousands of miles either side of Scotland. The eroded roots of that mountain chain are recognised today as the Highlands of Scotland. The formation of the Cairngorm Mountains was integral to this story.

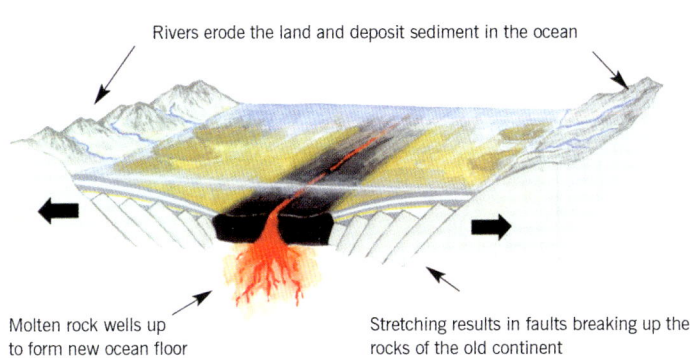

(a) Around 600 million years ago: the Iapetus Ocean was opening and 'Dalradian' sediments were being laid down.

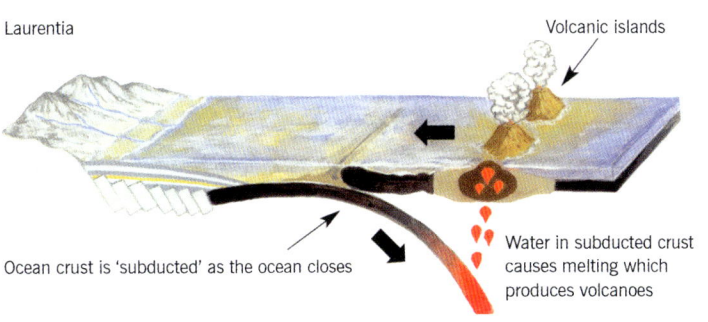

(b) Around 500 million years ago: the volcanic islands were on a collision course with Laurentia as the Iapetus Ocean was closing.

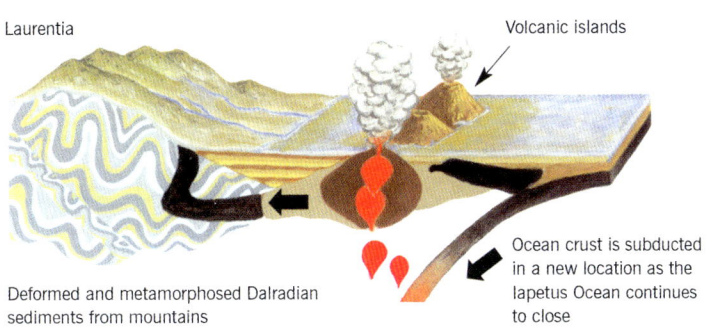

(c) Around 450 million years ago: collision of the volcanic islands and Laurentia formed a huge mountain chain.

rocks. The process of burial and compression heated up and fundamentally changed the rock from loose sands, muds, limestones and lavas into something much harder and more resistant to later erosion. This rock sequence is known to geologists as the Dalradian, named after one of the ancient kingdoms of Scotland.

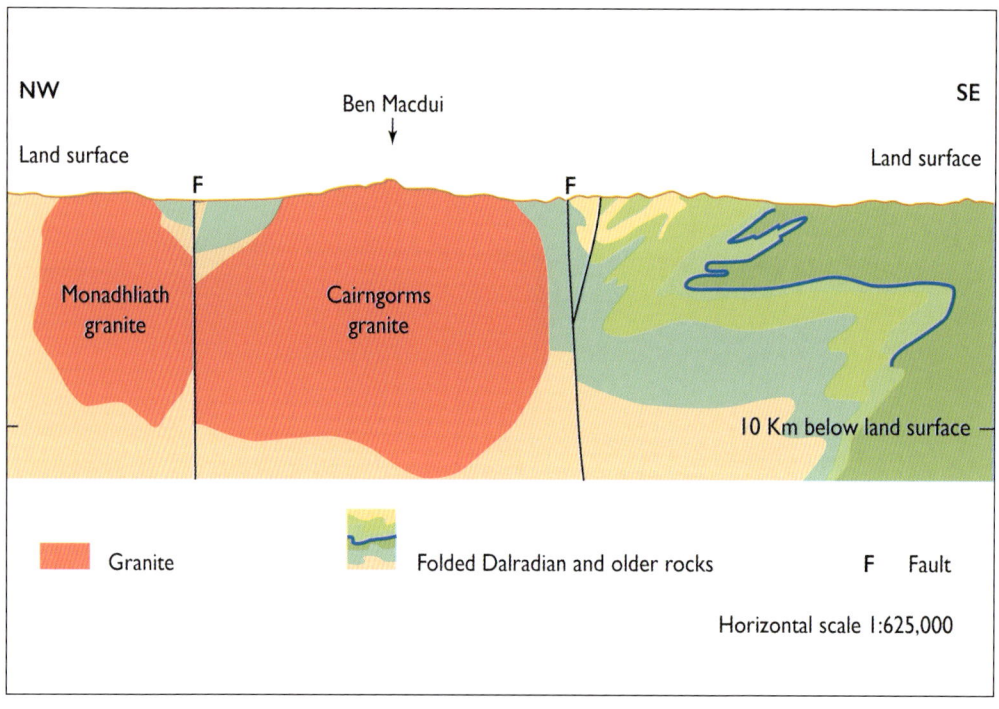

The Cairngorm granite

This illustration represents the aftermath of the continental collision. Like a piece of discarded carpet, the Dalradian strata were rucked up and bent double. As we'll hear more about in the next section, the temperatures and pressures generated by this collision were such that rock melted and huge quantities of granite magma were generated. As this molten rock was less dense than the surrounding strata, it started a slow ascent upwards through the Earth's crust. It pushed aside the newly folded sequence of Dalradian rocks, forming a series of granite intrusions.

The heat generated by this process was colossal. Molten rocks of granite composition were generated in huge quantities as a result of the head-on collision between continents. Melting took place deep in the Earth's crust, in fact close to the boundary with the underlying mantle. The molten rock that was to give rise to the Cairngorm granite ascended through the Earth's crust, eventually reaching a point between 4 and 7 kilometres beneath the Earth's surface as it existed at that time. In its ascent, the granite exploited a zone of weakness created by fault movements that dated back to much earlier times. There is no evidence of associated lavas reaching the surface, but it's impossible to prove either way as the evidence has been entirely removed by later erosion caused by ice, wind and water.

Many reservoirs of molten granite were formed at this time but all remained deeply buried beneath the land surface for many millions of years to come. It was only after the overlying kilometres of rocks were sliced away by powerful forces of erosion that the now solidified Cairngorm granite saw the light of day. It was, in the jargon, 'unroofed' around 380 million years ago during Middle Devonian times. From

From 380 million years ago onwards, the Cairngorm massif has stood proud. Dramatic rises in sea level during later geological times, notably during the Cretaceous Period some 180 million years ago, covered the rest of Scotland, but the Cairngorms stood above it all. This landscape was carved by ice in more recent times.

that date to this, the landscape of the Cairngorms has been evolving and changing in response to the climatic conditions.

This lengthy process of continental collision and its associated metamorphism created the Highlands of Scotland. But not quite as we would recognise them today, because the hills and glens were not sculpted by ice until more recent geological times.

This fist-sized specimen is typical of Cairngorm granite. It is fairly uniform in composition and consists predominantly of the most common rock-forming mineral on the face of the Earth – feldspar – with some quartz and a sprinkling of biotite mica. It also carries minor quantities of accessory minerals – zircon, apatite and magnetite. The granite massif was probably formed by a series of distinct pulses of granite, many of which vary slightly in both composition and texture. Fresh pulses of molten rock followed the previous ones in relatively quick succession.

The Cairngorms have a mineral named after them – a semi-precious variety of quartz known as smoky quartz, Cairngorm crystal or Cairngorm stone. Aluminium atoms are present in trace quantities in some quartz crystals, and natural radiation acted on these atoms, imparting the signature smoky appearance to individual crystals. Crystals of this mineral occur in veins within the granite and were very much prized in Victorian times. On 6 September 1850, Queen Victoria ascended Beinn a' Bhuird and noted in her diary that close to the summit 'we came upon a number of "cairngorms", which we all began picking up, and found some very pretty ones'.

Boulder drawn by John Clerk of Eldin. It shows the schistus cut by granite and then traversed by a later vein of red porphyry (a variety of granite). John Playfair, Hutton's principal biographer, said of the discovery 'The sight of objects which verified at once so many important conclusions in Dr Hutton's system, filled him with delight and the guides who accompanied him were convinced that it must be nothing less than the discovery of a vein of silver or gold, that could call forth such strong marks of joy and exultation.'

Glen Tilt – a historic site

Glen Tilt lies in the south-west of the Cairngorms area. The rocks laid bare in the bed of the River Tilt were the subject of detailed examination by Dr James Hutton, founder of modern geology and a leading figure of the Scottish Enlightenment. He visited this place in 1785 with 'his invaluable friend' John Clerk of Eldin. Hutton had just published his seminal work 'Theory of the Earth' in the newly established journal *Transactions of the Royal Society of Edinburgh*. The purpose of his forays into the field was to test his theories. Pre-eminent among his ideas was that the Earth acts as a 'heat engine' and that granite and all other igneous (volcanic) rocks were intruded in a liquid state and, demonstrably in this case, were not the oldest rocks of all, as was the accepted wisdom at the time. The host rocks into which the granite was introduced were the Dalradian schists, known to Hutton and colleagues as 'schistus'. What they found was that 'the granite broke and displaced the schistus (into which it was intruded) in every conceivable manner …' thus demonstrating that it was in a molten state when introduced into the host strata. This sounds like stating the obvious now, but it was ground-breaking in Hutton's day.

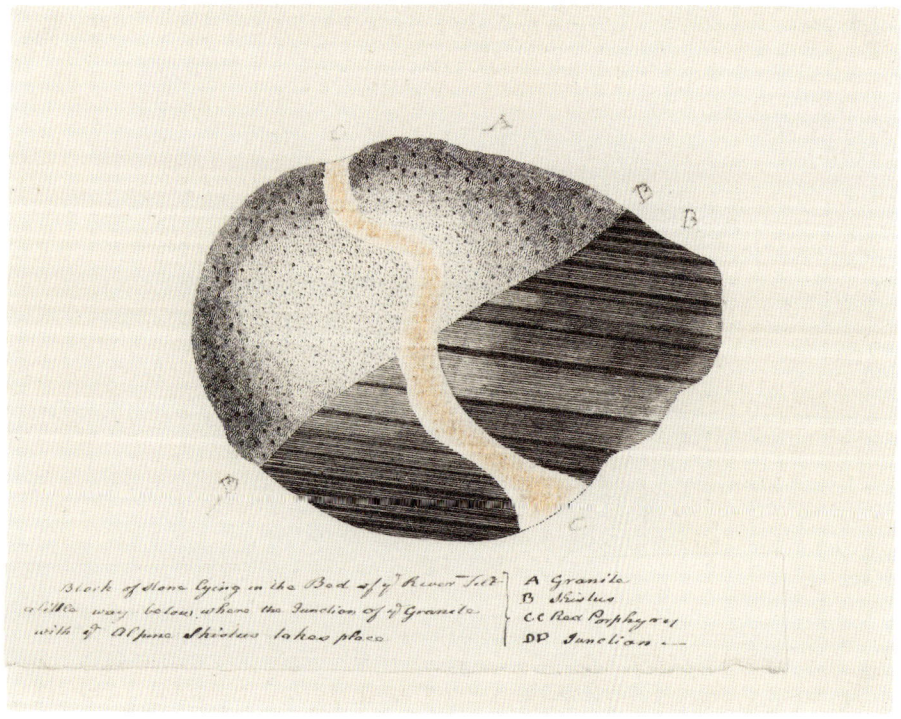

2
Devonian deserts

After continents collided over 420 million years ago, a new landmass was formed that lay just south of the Equator.

The erosive forces of wind and water got to work on the new mountain landscape soon after it was formed. In fact, erosion of the newly formed mountain range was spectacular in its rapidity. A conservative estimate of the amount of rock removed in a few million years

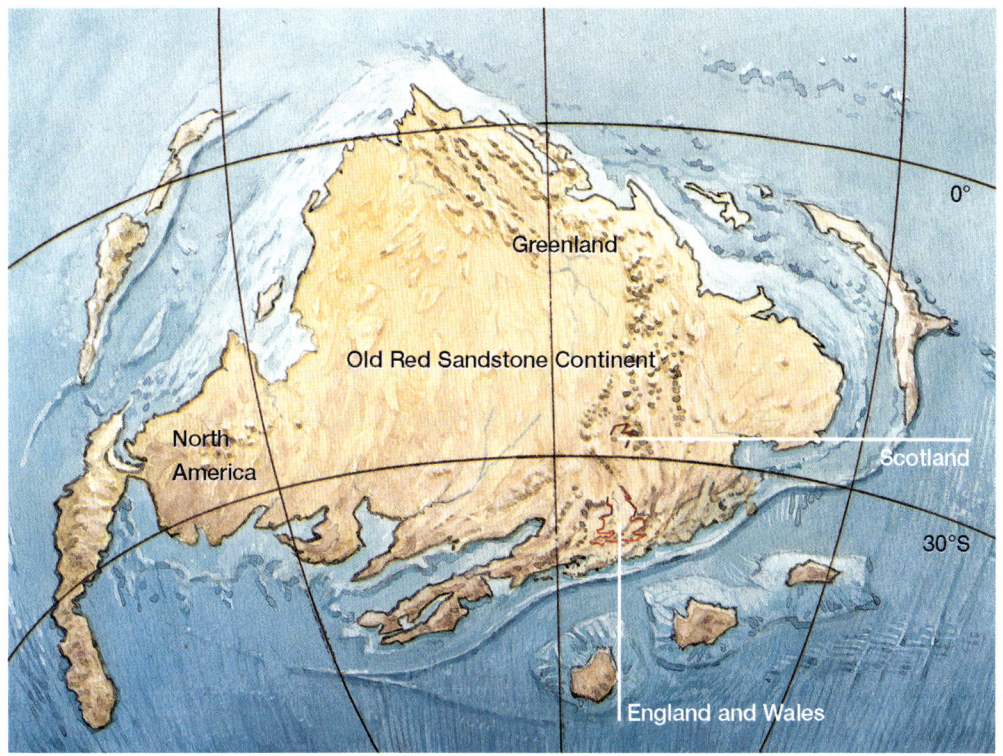

After the closure of the Iapetus Ocean, the land that was to become Scotland lay in a land-locked position, part of an arid and desolate landscape. This new continent also included North America, Greenland and most of continental Europe. Later convulsions in the Earth's mantle broke this landmass up into the more familiar geographic units that we recognise today.

Right. The Cairngorm granite formed deep within the Earth's crust in response to the processes that brought continents together in a violent collision. It was only much later when the overlying burden of Dalradian rocks was removed that the granite appeared at the surface.

Below. The landscape as it might have been during Devonian times.

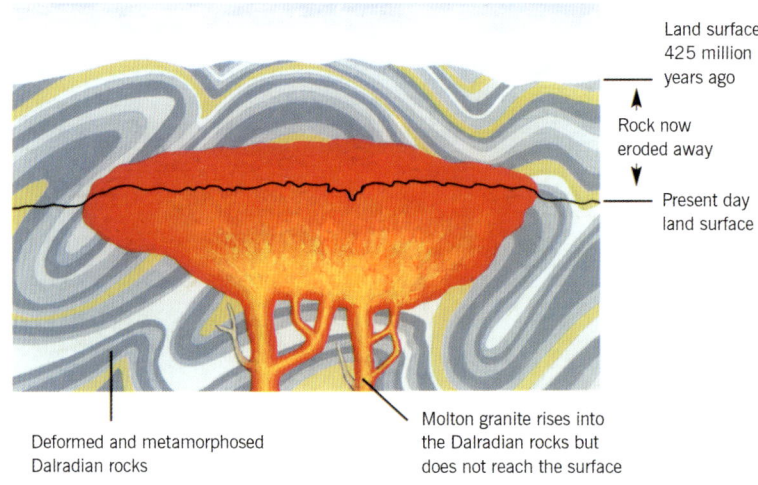

Land surface 425 million years ago

Rock now eroded away

Present day land surface

Deformed and metamorphosed Dalradian rocks

Molton granite rises into the Dalradian rocks but does not reach the surface

DEVONIAN DESERTS

is around 5 kilometres in thickness. That is extraordinarily quick work!

The landscape was dominated by the newly formed mountains. The wide plains were criss-crossed by streams that flowed from the high ground. The streams carried a burden of boulders, sands and muds to the lower ground where thick deposits of sediments built up in layers. It was in this arid environment that the rocks around the Tomintoul area were formed.

A steep-sided gorge near Tomintoul provides an excellent natural section through the rock strata. What remains today of these Devonian sediments (also known as Old Red Sandstone owing to their predominantly red colouration) has probably been greatly reduced from its original extent because of erosion by water, wind and ice.

Below. Ailnack Gorge, south of Tomintoul, is cut through Devonian layers of rock.

3
Mind the gap!

Since Devonian times, some 400 million years ago, many geological periods have come and gone without leaving a mark on the Cairngorm landscape. The land that was to become Scotland continued to drift north, through equatorial latitudes, and on into northern latitudes during this protracted period of time. For example, around 250 million years ago, a major re-arrangement took place when all the continents of the world came together to form one super-continent, but even this momentous event went entirely unrecorded in the Cairngorms.

Piecing together the geological history of Scotland requires an understanding of the whole landmass. There is no one place where the full story is told in the record of the rocks. The Western Isles are built from the very earliest rocks; the Edinburgh area largely records sediments and explosive volcanic episodes from the Carboniferous Period; and Skye is Scotland's 'Dinosaur Island' having yielded the remains of exotic and long extinct creatures from the Jurassic Period. A detailed knowledge of every area of the country is therefore necessary

This is a reconstruction of world geography as it existed some 250 million years ago. Driven by forces deep in the Earth's mantle, the continents came together to form one major landmass called Pangaea. 'Scotland' was located some 30° north of the Equator at this point and formed part of this huge landmass. At this time, desert conditions once again prevailed across Scotland. In the Elgin area, north of the Cairngorms, thick layers of fossilised sand dunes are preserved in the geological record that provide evidence for these conditions. But no such record of these or related events survives in the Cairngorms area.

The Cairngorm tors are among the finest in the country. The sub-tropical conditions of the late Cretaceous and into the Palaeogene Period 'rotted' the granite by deep sub-surface chemical weathering, but occasionally left a sound central core that was exposed as the surrounding rock fell away. Some tors can reach up to 20 metres in height.

The wide sweeping plateau area of the Cairngorms also formed before the ice.

to see the whole picture. Each area has blanks – gaps in the geological record when tumultuous events are writ large elsewhere in the country, but go completely unrecorded in immediately adjacent landscapes. Despite this, the collective record of the rocks in Scotland demonstrates much of the history of our planet.

In the Cairngorms, the 'blank' in the geological record exists between 400 million years ago and events in the late Cretaceous Period and into the Palaeogene. During these times around 65 million years ago, the climate was hot and humid and these sub-tropical conditions had the effect of deeply weathering the bedrock. Granite tors were one of the most striking landscape features generated by these conditions. They cap some of the peaks in the extensive Cairngorm plateau that also formed during at this time. The tors can be seen from some distance as distinctive 'pimples' on otherwise smooth hillsides.

From these balmy climes, the climate deteriorated and descent began into perhaps the most significant event to shape the landscapes of the Cairngorms – the Ice Age. That's the episode we'll explore next.

4
The big freeze

As the Neogene Period drew to a close around 3 million years ago, the climate started to cool quite dramatically. This gradual, but marked, climatic change had everything to do with the biggest planetary story of all – the Earth's elliptical orbit around the Sun. This eccentric orbit, as well as 'wobbles' of the north–south axis around which the planet spins, led to variable amounts of sunlight reaching the Earth's surface at different stages of these cycles.

When the Ice Age started to bite, the Cairngorms would have looked very similar to this present-day landscape in Antarctica. Great rivers of ice – glaciers – fed down from the higher ground, moulding and sculpting the landscape as they slowly ground their way downhill. Boulders, stones and other debris froze onto the underside of the ice, enhancing the abrasive power of the glaciers. The base of the glaciers acted like sandpaper, shaping and carving the bedrock as these frozen streams of ice inched from the higher ground towards the sea. It is this event that is largely responsible for the way the countryside looks today.

Above. As ice and snow accumulated on the north-facing slopes, this frozen mass moved downslope, under gravity, eroding large bites out of the Cairngorm plateau.

Left. These roadside rock exposures, or roches moutonnées, show the power of the ice to erode solid rock. The granites have been smoothed by the passage of the ice, forming an undulating surface known as a 'glacial pavement'.

The freezing conditions of the Ice Age became widespread across the country. A blanket of ice and snow extended south from the North Pole, imposing a harsh regime on the ecosystems that had previously become established. This event has had the most significant influence on shaping the landscape we recognise today. Mountains and glens, hills and straths owe their origins to these icy conditions.

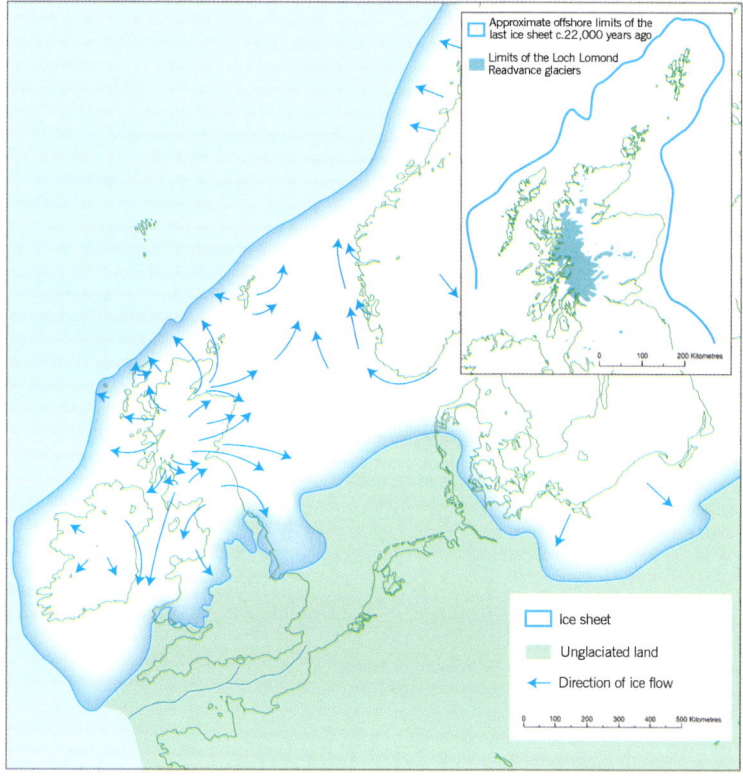

During phases of these cycles when the Earth was furthest from the Sun, temperatures fell. This was sufficient to tip the climate into a prolonged period of freezing conditions. We recognise this change as the onset of the Ice Age.

These cold spells lasted for around 100,000 years and were separated from the next big freeze by a period of warmer conditions, known as an inter-glacial. During these brief respites, lasting around 10,000 years, temperatures climbed and biodiversity recovered its foothold on many parts of the planet that had become frozen wastelands.

During the Ice Age, there were many advances and retreats of the ice as the climate changed. The level and extent of ice cover changed in step with the temperature. Because successive glaciations largely removed the evidence of previous episodes, the geological record of some Ice Age events in the Cairngorms is fragmentary and incomplete. But the later events are well recorded: for example, around 22,000 years ago the whole of Scotland was swathed in an icy blanket.

THE BIG FREEZE

Great gouges in the landscape, such as the Lairig Ghru mountain pass, were formed by moving ice. The sides of the glen have softened in recent times as debris flows, known as landslides, have developed.

Between around 30,000 and 15,000 years ago, even the very highest ground of the Cairngorms was overtopped by ice. Relatively fast-moving streams of ice cut great scars into the bedrock, such as at Glen Avon and, perhaps most spectacularly of all, at the Lairig Ghru mountain pass. In contrast, the ice that covered the high plateau was thin and slow-moving. As it was probably frozen to the rock surface, it didn't cut deep or modify this undulating surface to any great extent. This explains why the tors, which were initiated before the great freeze, survived largely intact.

Around 15,000 years ago, at the end of the last cold snap, the climate started to warm and pioneer grasses gained a toehold. All of these events can be reconstructed and dated by the analysis of pollen grains recovered from sediments that built up at the bottom of adjacent lochs and in accumulated peat deposits.

Around 12,900 years ago, the climate cooled once again with an average summer temperature of around 5°C. Small glaciers reappeared on the higher ground in the Cairngorms but this return to glacial conditions only lasted for 1,500 years. By 11,500 years ago, the last of these short-lived glaciers had melted and we entered the geological phase we currently enjoy – the Anthropocene. It was so named because of the effect that Man started to have on the face of the planet, its ecosystem and ultimately its climate.

Torrents of meltwater

On each occasion when the ice melted in response to a changing climate, deluges of icy water were liberated to flow across the landscape. Rivers and streams act like conveyor belts, transporting these water torrents, boulders, cobbles, sand and mud from the higher ground to lower flatter lands and onwards to the sea. They are powerful agents of change that have been scouring the landscape for millennia.

One of the sites that illustrates the volumes of meltwater involved lies to the north of the Cairngorms on the River Findhorn at Randolf's Leap. The 'boiling' torrents of water and their sediment load had the ability to act as a powerful rock-saw that sliced through the underlying bedrock.

The melting waters flowed more gently at other times. Huge braided streams and rivers flowed from the melting ice across the lower

As the ice melted, water was shed from the high ground in colossal quantities. The water and its sediment load acted as a rock-saw as it carved a slot into the underlying bedrock at Randolf's Leap.

Flood line 50 feet

THE BIG FREEZE

ground excavated by the glaciers. Great spreads of gravel and sand were deposited in a geological instant. Later erosion by rivers has cut through these deposits to reveal their extent.

Another feature of this de-glaciation or ice melting process was the creation of great ice-dammed lakes. The larger valley glaciers took longer to melt, blocking the escape route of water melting from adjacent ice fields. Small temporary areas of standing water resulted. Evidence for this is found in Glen Quoich and Glen Derry in the south and Lairig Ghru and Gleann Einich further to the north. (See diagram overleaf.)

The dynamic process of landscape change and renewal continues to this day.

Another spectacular, and accessible, place to see the effects of the torrents of water flowing from the melting ice is at the Burn o' Vat just to the west of Aboyne, Aberdeenshire. A deep, roughly circular gouge was cut into solid granite by rushing meltwaters from the

The River Feshie has cut through the layers of sand and gravel deposited by the meltwaters from the decaying ice sheets.

As small glaciers melted away due to rising temperatures, the larger ice streams remained in place. Water formed a pond between the two masses of ice, laying down layers of sediment characteristic of still waters.

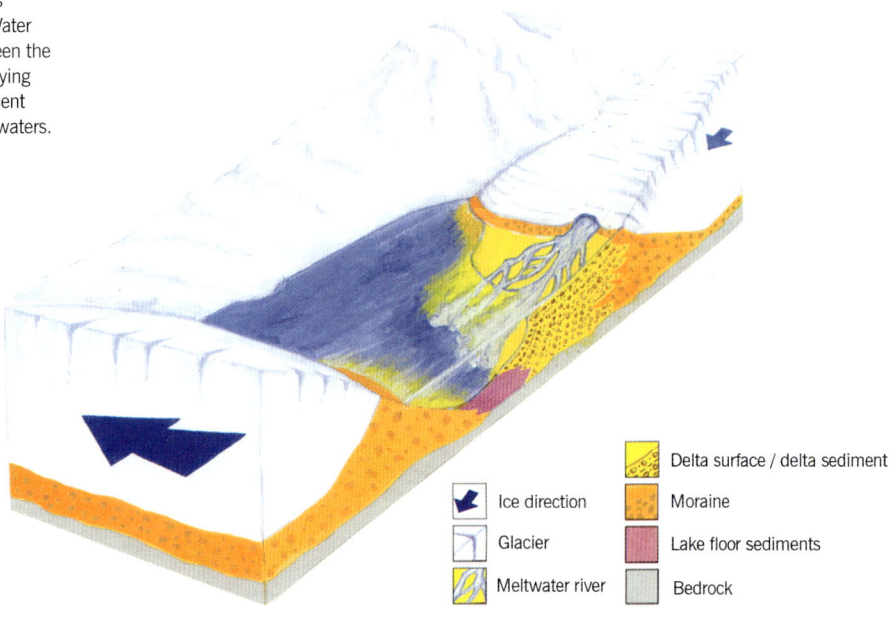

Opposite. This striking image of Glen Feshie, in benign mood, illustrates the ongoing nature of landscape evolution. The river has claimed a wide floodplain as it meanders down from the high Glen Feshie Forest. In times of spate, the river bursts its banks and follows a series of different paths. Previous river channels are clearly evident here, showing a landscape on the move. The bars and channels change position as a consequence of the dramatic floods, and the geography of the place is constantly being re-engineered.

adjacent high ground. Large boulders got trapped in a small depression and, over time, they circulated round and round creating a deeper and more pronounced pothole that is now some 15m deep and 18m in diameter at its widest point. After the ice melted and the torrents subsided, the stream that now passes through the depression diminished to a trickle. See 'Places to visit' for more details.

A landscape before people

Long after the mayhem created by continental collisions, grinding glaciers and torrents of meltwater, the landscape that resulted from these global geological influences has continued to evolve to this day. Not perhaps with the same drama and earth-shattering consequences, but change nonetheless. The ice only left the scene just over 10,000 years ago, leaving a raw and inhospitable landscape.

After the first pioneer species became established, the vegetation cover slowly increased in diversity. Grasses and sedges were first to appear, followed by shrubs and birch. We know the order of succession in which the plants appeared as the layers of peat that have built up during the last millennia have preserved the pollen grains shed by trees

and shrubs that formerly covered the area. These organic layers, with the oldest at the bottom of the peat bog and the youngest at the top, are like a biological time capsule that can be read. As a rough guide, a metre of peat takes about 1,000 years to accumulate.

By around 8,800 years ago, pine woodland covered much of the area up to a tree line about 800 metres above sea level. This massive extent of native woodland is known as the 'Great Wood of Caledon'. Today it is much diminished with isolated fragments flourishing in a few disconnected pockets across the Cairngorms. Re-establishing the woodland has proved difficult because of the omnipresent herbivores such as sheep and deer. But with more active management, larger areas have now been fenced off, allowing the seedlings to thrive. Glen Feshie is a particularly good example of a place where natural regeneration is now taking place. So perhaps the great Wood of Caledon may rise again!

In its heyday, this extensive area of woodland would have been home to a wide variety of animals that are now extinct in Scotland including wolves, wild boar, lynx and elk. There are moves to re-introduce some of these long-absent species, beavers and lynx particularly, to the ecosystems of Scotland.

Above. Burn o' Vat (described on pages 27 and 28).

Opposite. Natural woodland, such as this remnant of the Great Wood of Caledon in the Pass of Ryvoan, once covered much of the Cairngorms.

Below. Lynx, once found in the Great Wood of Caledon, but now extinct here.

5
The nature of the Cairngorms

Although some animals that once roamed the wide-open spaces of the Cairngorms are no longer present in this area, there are many other unusual species found here that make this a unique place. 'Unique' is a much over-used word that has lost some of its impact, but it certainly applies here. There is nowhere else in the UK where the winter conditions are so harsh and extreme. It's like a little piece of the Arctic has been transported south to Scotland! These tough conditions have produced a hardy range of plants and animals that can endure the cold conditions, occasionally interspersed with warmer days.

The Great Wood of Caledon extended across much of the landscape shortly after the end of the Ice Age. But within and between its remaining fragments and the extensive areas of the high plateau, a specialist and

This view of Glen Quaich is iconic of the Cairngorms. The heather-covered lower ground is studded by individual Scots pines and a more extensive wooded area that then gives way to deeply incised glens and high plateau typical of the area.

biodiverse community has established and continues to thrive.

The Cairngorms is quite simply the most important mountain area for plant and animal conservation in Britain. This is the view of Scottish Natural Heritage, the Scottish Government's statutory conservation agency. The range of designations (Site of Special Scientific Interest, Special Protection Area, National Nature Reserve and National Park) that have been applied to the Cairngorms Massif is wide, covering the conservation of mountain habitats, as well as a wide variety of insects and birds, including dotterel, snow bunting, golden eagle, merlin, peregrine and ptarmigan. A wide variety of mammals including otters, red deer, wildcat, badger, red squirrel and mountain hare all have sustainable populations that have thrived despite the extreme conditions. Vegetation cover includes areas of blanket bog, bog woodland and species-rich grassland. To find out more, visit www.snh.gov.uk/sitelink.

Let's look in more detail at some plants and animals that are characteristic of the Cairngorms.

- **Golden eagle** is iconic of the Scottish mountains, and the Cairngorms in particular. Its occasional languid wingbeats and unhurried progress across the sky live long in the memory of anyone lucky enough to see one of these magnificent birds on the move.

Golden eagle.

Scottish wildcat.

Dotterel.

- **Scottish wildcat** has become a *cause célèbre* in recent years. Its future in the wild is precariously balanced, although a concerted conservation effort is currently under way to save Britain's remaining population of native wildcats.

- **Dotterel** is a bird whose primary habitat is the high-level snow-fields. It is a summer visitor to the Cairngorms and is unusual in that females are more brightly coloured than their male counterparts.

Mountain hare.

Ptarmigan.

- **Mountain hare** do well in the mountains where there is a good heather cover. In snow, the hare's coat turns pure white to camouflage it from predators such as the golden eagle. This hare is in transition to its winter coat.

- **Ptarmigan** have made their home on the high Cairngorm plateau. During the summer the ptarmigan plumage blends into the surrounding rocks, while during the winter it turns a brilliant white, less visible in the snow to predators.

Nesting osprey.

Mossy saxifrage.

- **Ospreys** have been part of the Cairngorm ecosystem since the late 1950s, when a pair nested at Loch Garten. Since that time, this graceful fish-eating bird of prey has been ever-present during spring and summer, and now over 100 pairs are thought to populate the area. Over two million people have visited the RSPB Loch Garten centre to view the nest sites.

- **Mossy saxifrage** grows as mats on poor soils in the upland grasslands and scree slopes of the Cairngorms. The montane heaths are also rich in lichen species and other hardy plants such as dwarf heathers.

6
A natural playground

The Cairngorms have been a magnet for outdoor enthusiasts for more than a hundred years. Mountaineering, hill-walking, cycling, skiing and more recently snowboarding are some of the pursuits that are regularly enjoyed by large numbers of locals and visitors in this dramatic setting.

There are three long-established ski centres serving the area: at Cairngorm itself, Glenshee and the Lecht ski and activity centre.

For the more adventurous walkers, there is high ground, rugged peaks and unspoilt terrain aplenty in the Cairngorms. If we take Sir Hugh Munro of Lindertis' yardstick of peaks above 3,000 feet as representing a mountain, then there are around 20 of them to enjoy in the Cairngorm range.

Cairngorm has the highest and longest ski runs and there is also good cross-country skiing nearby across the plateau when conditions permit.

Right. The funicular railway runs from the high level car park on the northern slopes of Cairngorm to a point close to the summit. It allows access, albeit restricted, to a world of more extreme conditions in the winter months particularly. When the rest of the country is entirely free from snow, these north-facing slopes hold their snow cover for much longer, allowing skiers and snowboarders to indulge their passion. The Ptarmigan restaurant (the highest in the country), and the viewing terrace, shop and exhibition about the flora and fauna of the mountain, provide an additional attraction for those not equipped for sport!

Below. A walker high on the Cairngorm plateau.

Whisky on the rocks

Uisge beatha, or 'water of life', is a popular description of Scotland's national drink – whisky, the distillate of the simple ingredients of fermented grain, yeast and water. The rivers Spey and Dee cut a swathe through the Cairngorms and provide a plentiful supply of one of these key elements – pure water. This natural resource is used as part of the distillation process and also for cooling purposes in what is, after all, an industrial process.

Each whisky has a different taste and 'nose' that is recognisable to the connoisseur. Although the purity of the water is celebrated, the rivers and their tributaries bring an additional ingredient to the mix. The rocks over which the water flows impart a subtle natural signature to it that influences the flavour of the malt whisky. Limestone and sandstone give the flowing waters a neutral or alkaline character. In contrast, water in the form of rainfall falling on high ground built from granites or other crystalline rock will be rapidly shed downhill to lower ground with little opportunity of interacting with the

Lying within the Queen's Balmoral Estate, Lochnagar is a mountain appreciated by climbers, walkers, poets, painters – and geologists.

Deeside was formerly a place where illicit distilling of whisky was commonplace. On a site very close to the royal estate at Balmoral, James Roberston of Crathie built his third distillery on the banks of the Dee. (The first two were mysteriously burnt to the ground.) After a tour of the premises by Queen Victoria, Prince Albert and their family, the company were awarded a Royal Warrant. They must have served their royal visitors a good dram!

bedrock. In such circumstances, the water will remain acidic. The waters that feed the distilleries of the Cairngorms area fall into the latter category.

Many of the distilleries throughout Scotland have visitor centres that encourage tourists to share some of the secrets of the age-old craft of whisky distillation. Dalwhinnie at the extreme west edge of the Cairngorms, Royal Lochnagar near the Balmoral Estate and Glenlivet in the north-east of the National Park area are all places where this traditional industry is interpreted in an interesting and informative manner.

Cairngorms National Park

The Cairngorms was designated as a National Park in 2003. It is the UK's largest National Park, twice the size of the next largest. The National Park Authority play a key role in looking after the natural assets, such as landscape and wildlife, as well as regulating new developments within the boundaries of the park. They also have an important role in helping visitors to appreciate and enjoy the great outdoors in general and this special place in particular. To find out more, visit www.cairngorms.co.uk.

Cyclists enjoying the Cairngorms.

7
Places to visit

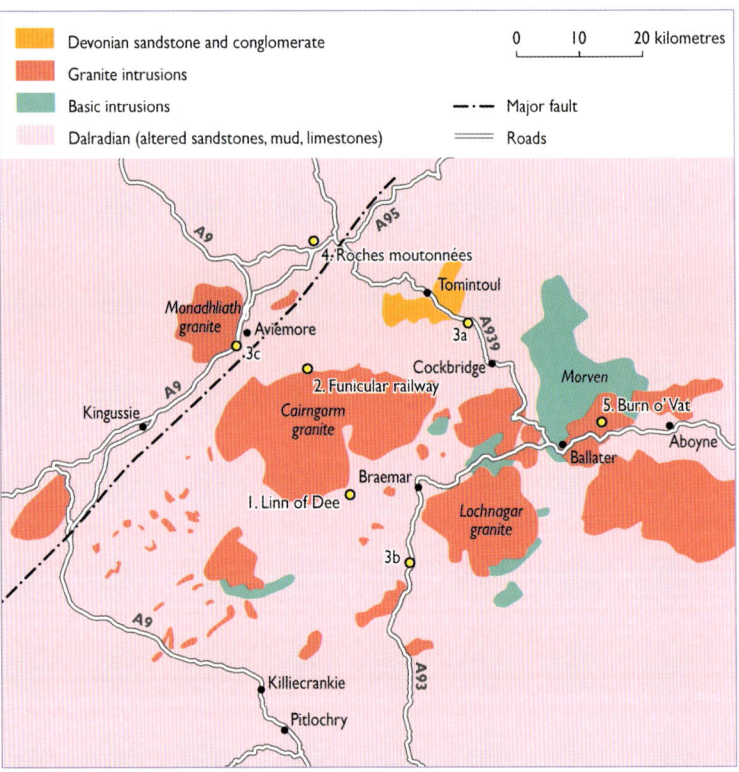

Map showing the locations of places to visit.

Weather can change in a heartbeat in the sub-Arctic environment of the Cairngorm plateau – from bracing at the beginning of a walk to extreme conditions a short time later. The places of interest described below are easily reached and safe to visit in all but the most extreme weather conditions. None of them are places where visitors may find themselves in unfamiliar, challenging terrain. There are many other sources of information available to inform the more adventurous visitor on all issues related to access to the countryside; see www.outdooraccess-scotland.com

Linn of Dee

The National Trust for Scotland (NTS), who own the Mar Lodge Estate, have a ranger service that offers outdoor events and walks – see Linn of Dee below. The OS Landranger Map series numbers 35, 36 and 43 are useful in navigating between these sites and also for planning your own routes.

1. **Linn of Dee** – the dramatic falls at the Linn of Dee are to be found a few kilometres west of Braemar. NTS have provided parking and toilet facilities. The River Dee, on its progress westwards, has carved a deep cleft through Dalradian rocks. This is also an access point to Glen Lui that runs north-westwards to Ben Macdui and the Lairig Ghru beyond. NTS run guided walks with qualified leaders for anyone who wants a real mountain experience. Booking ahead is essential for this trip.

2. **Cairngorm funicular railway** – the railway runs from a car park at the end of the access road to the Cairngorm ski resort up to the edge of the plateau summit. In winter, it is very much the skier and snowboarder's domain, but the upper station, by the Ptarmigan restaurant, also affords panoramic views across glaciated landscapes of the western part of the Cairngorms.

3. **Cairngorms from the roadside** – for those who want to appreciate the landscapes without pulling on their walking boots, the road network that runs around, and partly through, the Cairngorm massif is an excellent way to get a flavour of the place.

3a. The A939 runs between Cockbridge and Tomintoul, the highest village in the Highlands. It's one of the most famous routes in Scotland as it's often the first to be closed with the onset of wintry weather. Alpine in some aspects, it's a roller-coaster ride through some of the highest land the country has to offer. It also gives an excellent flavour of the landscapes of this unique upland area. For those who haven't the time or inclination to do serious walking,

it's the armchair way to see some of this rugged terrain up close (see photo overleaf).

The A93 winds through Glen Shee.

3b. The A93 from Bridge of Cally to Braemar also takes the motorist through some spectacular scenery, following the glaciated valley of Glen Shee northwards by way of the Clunie Water to Braemar. From there, the A93 runs eastwards to Aberdeen, never straying far from the banks of the River Dee.

3c. The A9 runs along the western flank of the higher ground and provides spectacular views of some of the main landscape features, such as the Northern Corries.

Above. Part of the route from Cockbridge to Tomintoul near the Lecht ski centre.

Right. Roches moutonnées information near Dulnain Bridge.

Opposite. Burn o' Vat signpost.

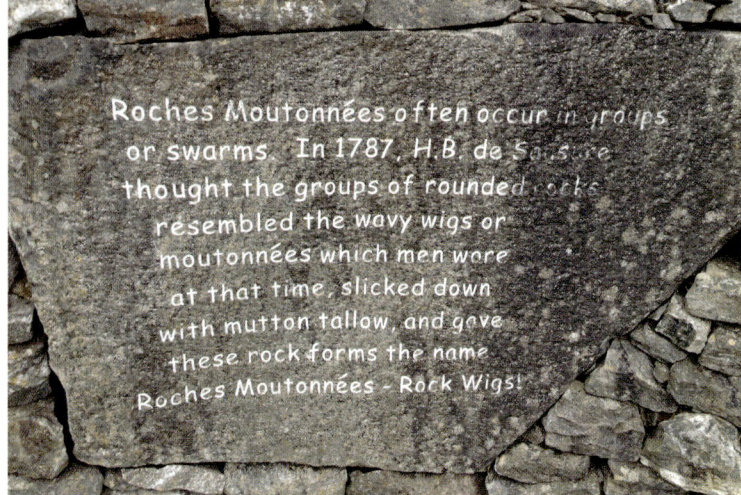

4. **Roches moutonnées, near Dulnain Bridge** – a brief description of this site is provided on page 23. These features of interest are located by the side of the road adjacent to the A95, near Dulnain Bridge just to the south of Grantown-on-Spey.

5. **Burn o' Vat** – a description of this site, and its formation, is provided on pages 27–28. This site forms part of the Muir of Dinnet National Nature Reserve. At Burn o' Vat, there is a visitor centre with toilets and ample parking.

Acknowledgements and picture credits

Thanks are due to Professor Stuart K. Monro OBE FRSE and Moira McKirdy MBE for their comments and suggestions on the various drafts of this book. I also thank Hugh Andrew, Andrew Simmons, Mairi Sutherland and Debs Warner from Birlinn for their support and direction. Mark Blackadder's book design is up to his usual high standard. Scottish Natural Heritage, in association with the British Geological Survey, published the 'Landscape Fashioned by Geology' series that was the precursor to the new 'Landscapes in Stone' titles. I thank them for their permission to use some of the original artwork and photography in this book. John Gordon, Rachel Wignall, Ness Brazier and Patricia Bruneau wrote the original text for *Cairngorms – A Landscape Fashioned by Geology* which influenced aspects of this book.

Picture Credits

Title page Jan Holm/Alamy Stock Photo; 6 Angus Findlay/CNPA; 11 drawn by Jim Lewis; 12 drawn by Craig Ellery; 14 drawn by Jim Lewis; 15 (upper) Lorne Gill/SNH; 15 (lower left) British Geological Survey; 15 (lower right) Lorne Gill/SNH; 16 reproduced with the permission of Sir Robert Clerk Bt of Penicuik; 17 drawn by Jim Lewis; 18 (upper) drawn by Craig Ellery; 18 (lower) Clare Hewitt; 18 Lorne Gill/SNH; 20 drawn by Jim Lewis; 21 (upper) Rachel Wignall; 21 (lower) John Gordon; 22 John Gordon; 23 (upper) Angus & Patricia Macdonald/Aerographica/SNH; 23 (lower) Alan McKirdy; 24 drawn by Jim Lewis; 25 John Gordon; 27 Lorne Gill/SNH; 28 Craig Ellery; 29 David Gowans/Alamy Stock Photo; 30 (upper) Alan McKirdy; 30 (lower) Lorne Gill/SNH; 31 (lower) Jasperimage; 32 Lorne Gill/SNH; 33 Martin Mecnarowski; 34 (upper) davehuntphotography; 34 (lower) Erni; 35 (upper) Sue Berry; 35 (lower) Ben Queenborough; 36 (upper) Brian Lasenby; 36 (lower) HHelene; 37 Angus Findlay/CNPA; 38 (upper) John Peter Photography/Alamy Stock Photo; 38 (lower) Keith Slater; 39 Simon Price/Alamy Stock Photo; 40 Scottish Viewpoint/Alamy Stock Photo; 41 Cairngorms Business Partnership; 43 Alan McKirdy; 44–5 Mark Caunt; 46 (upper) Harvepino; 46 (lower) Alan McKirdy; 47 Alan McKirdy